Hebe Three

A Numberline Lane book
by
Fiona and Nick Reynolds

Hebe Three lived in a very large house in Numberline Lane.

Her house had many rooms and lots of windows.

Hebe Three's bedroom was the largest bedroom in the house.

It had a huge three-poster bed in the middle of the room.

Hebe Three had decided to visit Jenny Ten to talk about a painting for her hallway.

Hebe Three always wore a hat that matched her shoes.

She went to her hat cupboard, opened the door and chose a bright yellow one.

"Perfect for a sunny day," she thought to herself.

Then Hebe Three went to her shoe cupboard.

She opened the door, and with a large crash all her shoes came tumbling out of the cupboard and onto the floor.

Hebe Three looked very carefully but she could not find a single matching pair.

"Oh dear, oh dear! What a muddle!" said Hebe Three, "How will I ever be able to sort out all these shoes?"

As Hebe Three was thinking, there was a knock at the door.

It was Walter One.

He had come round to have a swim in Hebe Three's swimming pool.

"Hello, Hebe Three," said Walter One.

"Hello, Walter One," said Hebe Three.

"I wonder if you could help me?" said Hebe Three, "My shoes are in a terrible muddle and I don't know how to sort them out!"

Walter One had a short think.

He suggested that Hebe Three should put all the left shoes in one cupboard, and all the right shoes in another cupboard.

With that, Walter One went off to the swimming pool.

"I don't think that is a very helpful idea," thought Hebe Three, "I still wouldn't be able to find a matching pair."

Then there was another knock at the door "Knock–knock, knock–knock".

It was Nora Four.

She had come round to give Hebe Three a new scarf that she had been knitting.

Hebe Three explained her problem to Nora Four.

"Well I think you should sort out your shoes by putting your comfy slippers in one cupboard and all your other shoes in another."

With that, Nora Four left.

Hebe Three thought about this for a moment, but decided that this still wouldn't help her to find a matching pair of shoes to go with her yellow hat.

So Hebe Three put on two shoes (that didn't match) and went to see Jenny Ten.

"Knock, knock, knock!" went Hebe Three on the door.

"Hello, Jenny Ten," said Hebe Three.

"Hello, Hebe Three," said Jenny Ten.

"Do come in. I have just finished the painting for your hallway. Would you like to see it?"

"Yes please," said Hebe Three.

As they were looking at the painting, Hebe Three explained her problem with sorting out her shoes.

"Well I would always sort everything out into colours," said Jenny Ten.

Hebe Three thought about this for a while and started to smile a very wide smile.

"What a wonderful idea!" she said, "I will go home and sort out my shoes into colours straight away."

That is exactly what she did.

When she had finished all her sorting, she went to the yellow shoe cupboard and found her favourite yellow shoes — a matching pair!